Game Theory

A Guide to Game Theory, Strategy, Economics, and Success!

Table of Contents

Introduction

Thank you for taking the time to pick up this book about game theory!

For game theorists, a game is the interaction between two or more people wherein the payoff for each person is affected by the actions and decisions of other participants.

Game theory can be used to assess and decide upon the best decision that should be made in a variety of situations. As you will discover in the following chapters, it can be used to improve outcomes in board games, psychology, politics, business, and more!

By studying, understanding, and implementing some game theory strategies into your life, you can greatly improve outcomes in all areas. This can result in better relationships, improved health, and more success at work and in business.

Once again, thanks for choosing this book. I hope that you enjoy learning about the fascinating science of game theory, and its many facets and applications!

Chapter 1 - Learning the Basics of Game theory

What is game theory?

Game theory involves studying the mathematical models or systems of cooperation or conflict between rational decision-makers. The mathematical models studied in game theory allow one to identify which decisions should be made, while considering the possible action that will be taken by other participants of the game. It aims to analyze every possible decision that other participants can take and how it will affect you, which in turn can help you choose the action brings the best benefit or outcome.

Before game theory was applied to different fields of study, it was first applied to zero-sum games. However, the exploration of this concept by John von Neumann and other experts turned it into a theory that could also be used in other fields.

Today, it's considered as the science of making logical decisions for humans, animals, and even computers.

How did this theory develop?

The first discussion on what could possibly be considered as game theory started from the letter of Charles Waldegrave, an active Jacobite and British diplomat in 1713. His letter contained a strategy on how to minimize one's loss even if they were faced with the worst case scenario, while playing the two-player version of Le Her, a classic French card game.

Game theory's influence in economics started with James Madison, one of the founding fathers of the United States. He made a game theory-like analysis on the different ways in which states can behave depending on the system of taxation that will be used.

Game theory started to exist as a unique field of study in 1928, when John von Neumann published a paper that made use of a mathematical theorem in applying his main idea. The theorem he used was also applicable to mathematical economics, which somehow explains why game theory and economics are related.

A huge development in game theory was observed in the 1950s, as this is when certain concepts such as the Nash equilibrium, Shapley value, different types of games, and the application of game theory in other fields of study such as political science and philosophy all happened.

Even now, game theory is still being developed. Experts are getting awards for providing substantial improvements and additions to the theory, helping it to keep up with the change of times.

What is a game in game theory?

For game theorists, a game is the interaction between two or more people wherein the payoff for each person is affected by the actions and decisions of other participants.

One element present in a game is rules. No matter what type of game is being described, the following rules are always observed:

1. Point of the game – any game has its stakes and it specifies the condition that should be met to determine who wins.

2. Participants – a game cannot be played without determining who are qualified to become participants or "players".

3. Field of play – this rule specifies the actions that are required for players, which actions they can take, and what actions are not allowed.

4. Strategy – refers to the sets of actions that one can take in order to maximize their chances of winning the game.

In order for one to participate in a game, they should be aware of the game's purpose, who are its participants, and the actions that can be used to win it. Knowing these three rules is enough to say that one is playing the game. The fourth rule, though, is not a real requirement. One can play a game without applying a strategy. However, it would be in their best interest to have a few so that they can influence how the game turns out.

Types of games in game theory

Games in game theory can be placed in one of each of these categorizations, depending on their nature. Each category has two opposite poles.

Cooperative versus non-cooperative

A game is considered to be cooperative if it's possible for the participants to form an alliance or coalition in order to maximize the benefits that they can get. Participants of a cooperative game can make use of binding agreements so that one can get "fair compensation" for their participation. The binding agreement also specifies how much each participant should contribute to the alliance so that everyone gives equally as well as receives benefits equally. Cooperative games make use of the Shapley value, which is the method used to divide the cost or gains among the participants depending on the value of their contribution.

On the other hand, non-cooperative games are situations where participants are not allowed to form coalitions, or can only do so with the use of force or threat. Participants of a non-cooperative game try to determine which course of action will give them the most benefit, regardless of what other participants decide to do.

Non-cooperative games make use of the Nash equilibrium to determine the best course of action.

Perfect versus imperfect information

Games can also be categorized depending on the information that is available to its participants.

A game is considered to be under the perfect information type if each player is able to see the move of the other participants. Chess is the perfect example for this type of game categorization.

On the other hand, interactions or situations wherein the actions of each participant are hidden from others fall under imperfect information games. Good examples for this type of game are card games such as poker. Even if there are "tells" that can be observed from one's facial expression, these are unclear indicators of the hand that the other player has got.

Sequential versus simultaneous

Sequential or dynamic games are situations where participants take turns to make an action, giving them an idea about the action taken by the previous participant. Many board games fit into this category.

On the other hand, simultaneous games are situations wherein participants move at the same time.

One thing to take note about this duality is that both game types do not necessarily require one to receive full information like what's available in the perfect information game type. No matter how little the information available to them, it is usually enough to influence the decisions they will make.

Symmetric versus asymmetric

The symmetry mentioned in these game types refer to the goals that each participant has.

A symmetric game refers to any situation wherein the outcome desired by each participant is the same. Many recreational games require the players to have one goal, as these games usually have single win conditions. The game of chess is of this game type, as the ultimate goal for winning to game is to capture the king or place it in a position where it cannot escape being captured.

On the other hand, asymmetric games are situations in which the participants have different goals, and the results of the interaction depend on each of their perspectives. For example, one business owner may want to win a contract in order to expand their company, while another business owner wants to win a contract for the purpose of mitigating the losses that the company is currently experiencing.

Zero-sum versus non-zero-sum

Zero-sum games are situations wherein the gain of one player translates into a loss for another participant. Think of a basketball game wherein one emerges as the winner while the other is the loser. This is a good example of a zero-sum game.

On the other hand, non-zero-sum games are situations wherein the action of one participant can bring benefit to multiple participants. A good example would be economic interactions wherein one participant in a coalition does what they can to increase the market, which is not only beneficial for them but for the other members of the group as well.

Chapter 2 - Strategies in Game Theory

It was emphasized in the previous chapter that the outcome of a game can be influenced by the strategies used by the players. Game theorists rely on different concepts so that they can formulate the best strategies.

This chapter will discuss more about strategy, and which elements should be considered when deciding upon an effective approach.

Importance of payoffs

Every game is played by participants who expect to receive some form of payoff for their actions, or for winning the whole game.

Payoff is often expressed as a unit of money, time, or anything that gives value or utility to a person. The player often ranks the utility or payoff that they can receive in order of importance to them, or assign a numerical value to it to denote its importance. In most cases, people prefer to receive utility when the other option is settling for less value. This only means that in any strategic situation modelled as a game, a person will often choose a course of action that can bring them more utility. In a way, it's a good way to learn more about the possible strategy that the player can use.

Use of the payoff matrix

Every person or company acts in order to get some form of benefit. However, they should carefully think of the actions that they will take so that they can get the best possible result out of it. To do this, they make use of the payoff matrix.

This diagram aims to determine what they can get if they take (or do not take) a certain action, and compare it to that of another company in their industry, specifically one that took the same course of action. This will inform them of the best action to take and where they can most benefit from regardless of what the other party does.

Extensive form

Another diagram that is highly important for game theorists aiming to arrive at the best decision possible is the extensive form, or the game tree.

This diagram is mostly used on games where players take turns in making a move, and there are different scenarios that one can make depending on the previous move. For example, if Player 1 makes move A, game theorists try to think of the best countermoves for it and place them underneath it. They do the same for all other possible moves that Player 1 can do. They can also assign numerical values for each move or countermove so that the party involved will know which action can give them the best payoff. In a sense, it makes use of the payoff matrix, but in a more extensive manner.

What is a dominant strategy?

Since every player acts in order to get the best possible payoff, they need to learn which courses of action always bring the best result, even if they have no idea what the action of the other party will be. This is what a dominant strategy is.

Knowing what the dominant strategy is (or if the game even has one), is important because it already dictates what the other player's action might be. Since they get the most benefit out of that strategy or action, it's highly possible that they will not deviate from it. This somehow makes the move of the other party more predictable, allowing them to take countermeasures so that their competitor will not be able to follow the dominant strategy.

Dominant strategies can be classified as either strongly dominant or weakly dominant. A strategy is considered to be strongly dominant if given all options, one will be able to get the highest payoff by choosing that course of action. On the other hand, a strategy is weakly dominant if the payoff that one can get from it is either as high as the other strategies regardless of the other player's choice, or if it is higher if compared to some of the strategies that can be used by the opposing party.

If possible, it is highly advised to throw in a mix of strongly dominant and weakly dominant strategies. While the payoff of the player who makes use of this mixed strategy may be affected and will not be that high if it is compared to just using a strongly dominant strategy, it can also affect the payoffs that the other player can get. In a sense, their own payoffs can look relatively larger compared to those of others.

What is an iterated dominant strategy?
There are games or situations wherein the player has no dominant strategy. Imagine that you are in a game of chess and you will have to face a player who doesn't follow a pattern of opening or endgame moves. If this is the case, how are you going to make a rational decision and choose which course of action should be taken?

Even if this is the case, drawing lots or even picking something at random is out of the question. For this kind of situation, game theorists make use of a solution that they call iterated dominance.

Iterated dominance is considered as the refined version of the dominant strategy. This is because this concept aims to further narrow down one's prediction of the game by eliminating choices that will most likely not be played by one participant. Thus, even if a player doesn't seem to have a strategy, one can still have a good idea of how the other person will act so that they can get the best possible payoff.

To further understand this, let's take a look at an example (Player 1 can make moves A, B, and C, while Player 2 makes moves 1, 2, and 3):

	Move 1	Move 2	Move 3
Move A	1, 10	3, 20	40, 0
Move B	10, 20	50, -10	6, 0
Move C	2, 20	4, 40	10, 0

In the example above, you can see the possible payoff for each player depending on the action that they will take. From the figures above, there is no dominant strategy that can be applied. It's also easy to spot that regardless of the action that Player 1 will take, Player 2 will surely not play move 3; after all, he gets no benefit from it. With this in mind, Player 1 can conclude that move 1 can be a strictly dominant option for the opponent, as it is slightly better than 2 and far more beneficial than move 3.

Even if this game has 3 possible moves for Player 2, using this method allows them to narrow down the possible actions that the latter can take by simply working backwards and considering which option can give them the least benefit. Once it's done, Player 1 is now able to anticipate what the opponent will do, as they consider one of the options to be "eliminated" from their opponent's possible strategy. This also defines what Player 1's action should be, as it's highly possible that move 1 will be chosen no matter what he does. To maximize his payoff then, Player 1 should choose move B.

That is how iterated dominance works. One should determine which strategy is dominated by others. Once found, it is eliminated. Simply repeat the process depending on the number of choices that a player can make until only one remains. In this context, a final prediction can be made as to which possible moves each party will take. Always assume that the other player can also anticipate what you have in mind, as this will allow them to strictly identify which strategy is dominated by the other.

One possible question regarding the use of this method lies in whether there's a need to eliminate choices in the right order. While it is ideal to do it in a way wherein the option that provides the least benefit gets removed first, this process can be done properly as long as the steps in eliminating options are followed. Simply put, there is no correct order of elimination. At the end of this process, the result will always be the same.

One point that should be remembered about this technique is that it is only advised to be done if there are other options available to the opponent which provide a higher payoff. This technique can be difficult to implement if the other options offer the same or higher payoff. It is only by eliminating the strictly dominated strategies that can one arrive at the best possible outcome in a game.

Nash equilibrium

It is also possible to encounter a situation that does not have a strongly dominant or iterated dominant solution. For these kinds of situation, game theorists make use of the so-called Nash equilibrium.

This concept is named after John Nash. He developed this concept while studying at Princeton for his doctorate degree in Mathematics. Although he was considered mad because of his belief that "aliens" were sending him coded messages in the front page of a newspaper, he was able to recover from it after spending many years in and out of different mental institutions. In 1994, he won a Nobel Prize in economics because of this concept.

Nash equilibrium occurs when the strategy of each player is the best possible response to those of their opponents and each player is acting optimally so that they can get the best payoff, as doing otherwise can give the opponent an advantage while they get less.

How do you find the Nash equilibrium?

The best way to identify which course of action is essentially a Nash equilibrium is to first know which outcomes are not, as anything that remains in the payoff matrix is considered to be as such.

A shortcut that can make it easier for you to identify the Nash equilibrium (this may not be applicable in some situations) is to identify the best responses of each player for every possible action that the other player can make. Underlining or marking the best response in the payoff matrix allows you to easily see which course of action is beneficial for each player. Any action wherein both payoffs are highlighted is the Nash equilibrium for that game.

Here are the steps that should be done to make this possible:

- First, assume that Player 2 plays move X. Determine which action will give the best payoff for player 1. Since he can get the payoff of 5 by choosing action A, underline that number.

- Assume that player 2 takes the other 2 options. Responding to move Y, since Player 1 gets the payoff of 3 for action C, he should make that move first. Finally, for move Z, player 1 should take action B to get a payoff of 4.

- Do the first three steps, this time under the assumption that Player 1 will make a move and Player 2 will be the one to check on his payoffs.

Here is a diagram where these steps can be applied (Moves A, B, and C belong to Player 1, while Moves X, Y, and Z belong to Player 2. First payoff is for Player 1):

	Move X	Move Y	Move Z
Move A	**5**, 1	2, 0	2, **2**
Move B	0, 4	1, **5**	**4**, **5**
Move C	2, 4	**3**, **6**	1, 0

For this diagram, one can see that the actions where both payoffs are underlined are (C, Y) and (B, Z). If player 1 chooses to play either move B and C to maximize his payoff, player 2 will also have the best payoff by choosing Z and Y respectively. Deviating from these options can cause a reduction to one's payoff while the other player gets the upper hand. For example, if player 1 chooses move A to gain 5, player 2 could respond by making move Z and reduce the former's payoff to 2.

Is there a difference between a dominant strategy and Nash equilibrium?

If you're to check on the definition of a dominant strategy and that of the Nash equilibrium, it can be easy to get confused because both of these concepts aim to determine which course of action provides the best benefit.

There is some form of overlap between these two concepts. Since a dominant strategy is considered to be the strongest solution in a game, we can somehow say that that strategy can also be considered as Nash equilibrium; after all, one can experience a loss or a much lower payoff if one deviates from that strategy. However, solutions that have found Nash equilibrium are not necessarily dominant strategies. This is also true for iterated strategies.

To simplify, all dominant or iterated dominant strategies are Nash equilibrium. On the contrary, Nash equilibrium solutions are not necessarily dominant or iterated dominant strategies.

It is because of this fact that game theorists are more focused on determining the Nash equilibrium in the game. Once they are able to find this, only then can they evaluate if the Nash equilibrium that they have found is a dominant strategy or not.

The pub manager's game

A good example of a game that has both Nash equilibrium and a dominant strategy is the pub manager's game.

In this game, two managers of a pub/bar are simultaneously considering giving their customers a special offer. The purpose of giving a special offer is not to compete with the other pub; rather, the aim is to increase their sales and improve their customer base. However, there is still the possibility that the other pub will make the same action. What should they do?

Here is a diagram that depicts the payoff of Pub A and Pub B respectively depending on which action they will take.

	Make the special offer	No offer
Make the special offer	10, 14	18, 6
No offer	4, 20	7, 8

Looking at the diagram above, it's easy to see that each party has a dominant strategy—which is to make the special offer. This is because the row of special offer shows Pub A that he can increase his payoff to 10 or 18, depending on what the other pub manager does. This is a significant number if compared to the payoff of 4 and 7 if they choose to not give the offer. The same is true for pub B; getting a payoff of 14 and 20 regardless of what

Pub A does if they give the offer is significantly better than the 4 and 7 if they chose not to go through with it. The significant difference in the numbers is the best outcomes for each pub manager, and is therefore the Nash equilibrium for this kind of situation.

Friends or enemies?

If it is possible for games to have both a dominant strategy and Nash equilibrium, there are also situations wherein there is no Nash equilibrium. A good game that provides this situation is friends or enemies.

For this game, let's assume that Raven is hiding from Logan, while Logan wants the opportunity to see Raven and be with her. Raven is planning to go out that night, and she can either go to a restaurant or to the club. Logan, having known this, also plans to go out. Since Raven doesn't want to see him and considers this as her preferred payoff, a choice where both can benefit is possible. If Raven sees him in the restaurant, she will go to the club. Logan, on the other hand, would want to switch his original choice to the club if Raven does this. This will circle back to Raven choosing the alternative, and the process will be repeated until he stops or there is no more opportunity to follow her.

Here is a diagram that will help explain this situation (Raven's outcomes are in the rows while Logan's are in the columns):

	Restaurant	Club
Restaurant	-1, 1	1, 0
Club	1, 0	-1, 1

These situations wherein one party always wants to deviate, resulting in having no Nash equilibrium is known as a game of pure conflict. Situations like this are difficult to predict because of the first mover's disadvantage. One way of dealing with this is to mix strategies so that Player 2 cannot predict where the other participant will be going.

Backward induction

One method that is usually used in extensive form matrices is backward induction.

This process involves going back to terminal nodes in the matrix to see which decision is Nash equilibrium, based on the payoff for each combination of decisions.

To better understand this concept, take a look at this diagram.

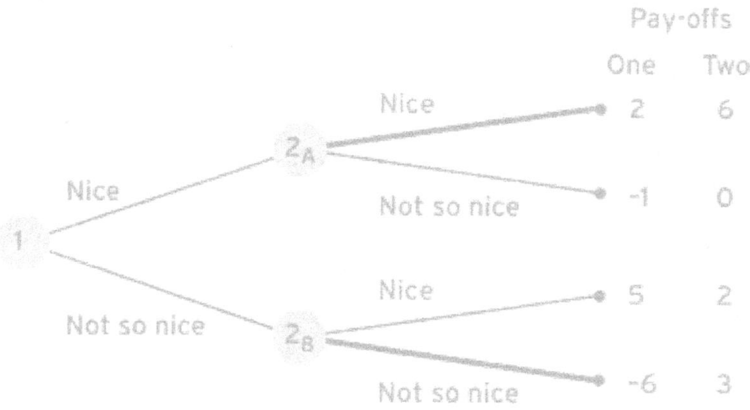

This is an extensive form for the game nice or not so nice. In this game, Player 1 should choose between being nice towards player 2, or to do otherwise. After Player 1 moves, Player 2 gets to act, reacting based on the payoff that he can get from it. In this diagram, it is shown that the terminal node is when Player 1 showed a nice or not so nice attitude to the other, which then branched out to the actions that Player 2 could make, along with the payoff for each.

For this example, go back to decision 2A, assuming that player 1 played nice. Player 2 now has the option to get a payoff of 6 for being nice, or reduce the opposing player's payoff by not being nice to them. Since 6 is greater than 0, Player 2's best action should be to act nice as well. On the other hand, it's possible to do backward induction to decision 2B and weigh the payoffs. Since not being nice gives the payoff of 3 compared to its opposite that brings only 2, the best action for player 2 is to react in a not so nice manner.

Using this method is necessary for both sides involved in the game. Player 1 can also plot the payoff for each decision in this manner. If the game is repeated or should the player decide to make the form more extensive and follow on the decision made by Player 2, he can see which action can give him more benefit and find out that such is the Nash equilibrium for the game.

Chapter 3 – Applications of Game Theory

It was mentioned earlier that game theory can be applied in other fields of study. Here, we will enumerate some of the fields that make use of game theory concepts, and discuss how those are applied.

Economics

This field of study heavily uses game theory. The competition between businesses can be considered as a game, and the payoff is getting a share of customers who are looking for products that they are selling.

The existence of any business entity can only continue if they regularly monitor what the other companies in their industry are doing. Every company in the market is considered as an entity that is capable of thinking and acting based on their interests. Failing to monitor the actions of other companies and the market in general can cause a company's downfall.

Some areas in economics and business that make use of game theory are as follows:

- **Price leadership** - since companies are always in the know when it comes to what their competitors are doing, they are aware of important details such as their prices for similar goods or services. One company can set a benchmark price on a certain service that others could follow. For example, if one airline company thought of charging extra for checked baggage that other airlines are not yet doing, they can set their own price for that service. This, in turn, may cause other companies to follow this benchmark price and also adopt this service.

- **Use of non-price methods** - although the price of a product or service can be a strong motivator for getting consumers to spend money, companies can also engage in techniques that will not change the price but still entice customers to choose them. For example, companies that sell high-priced items can highlight the advantage of their product to justify the price. This will enable consumers to know what they are buying and influence them to choose a specific brand; after all, the price is justified because of the product's good specs.

Psychology and social behavior

Many human interactions involve the use of game theory. Even the simple choice of which movie to see with your friend is influenced by game theory. Following this example, if a friend expressed that she wants to see Movie X but you want to see Movie Y, you can improve your relationship with that person if you choose to see Movie X with them rather than see your preferred movie alone. Any interaction where 2 or more people who can make a decision are involved, will use game theory, and the payoff is almost always improving one's relationship with that person or group of people.

The adoption of certain traits is also influenced by game theory. If a personality trait is favorable to a group of people, it's highly possible that they will develop that trait in order to gain acceptance.

The same is also true for evolutionary traits in animals. Take altruism as an example. Some animals make distress calls when they encounter a predator so that everyone can get to safety. While they can adopt the behavior of not making a distress call so that they will not jeopardize their own safety, it's in their best interest to do so. This is because if the predator is successful in preying on other members of their species while a few remain safe, there is a huge possibility that their population can be totally wiped out. If adopting that trait will ensure the survival of many members, it will eventually be passed to other members, making that behavior part of the future generation's instincts.

For psychology in the field of work, one concept in game theory is to know what the employee considers as a payoff that has high utility to them. For human resource practitioners to find a good candidate that will stay with the company or to motivate their employees to do what is needed, they should know what makes the employee work. If the practitioner stumbles upon an applicant that has the skills needed by the company, they can target whatever it is that the latter considers as the most important payoff. If the potential candidate considers the best payoff to be the opportunity for learning and leading a group, the company should be able to tailor their message so that the applicant will be enticed to accept the job offer.

The battle of the sexes game

One game that is closely related to the situation of two people selecting where they should go, is the battle of the sexes.

This game has two participants, and it is very similar to the situation mentioned above. The participants want to meet at a later time to hang out. However, each has their own preferences. One prefers to go to the bar, while the other wants to attend a party.

Many situations in real-life interactions are considered as simultaneous games such as this, and there is a huge possibility that each of them can choose the opposite of where the other person will be going. For more intimate relationships, interactions of this kind are critical; after all, there is an expectation that the other person should know what they want. However, even people in an intimate relationship can commit this coordination failure.

For this situation, it's easy to see that each person choosing what they like or what the other person likes are the worst outcomes. Person A wouldn't want to go to the bar or party without B, and vice versa. The problem though is how can it be possible for both parties to go to the same location even if they can't fully communicate as to where they should go? Game theory suggests that social interactions which appear to be a simultaneous game can sometimes look like a sequential game by pre-committing to their preferred location.

For example, if Person B takes considerable time in telling stories about the people at the party, Person A should pick up on the signal and choose to make the same decision. It's as if one is taking the first move so that the other will learn what their partner wants without explicitly telling what that is. Doing this improves the possibility that both will go to the same place later in the day.

Hawk and dove game
A game that is closely related to the idea of survival is the hawk and dove game.

In this game, it is assumed that there are two kinds of animals with different animal (or even human) behaviors. One participant is the hawk, which is known for their aggressiveness and tendency for fighting. The other participant is the dove, which has a more peaceful nature and prefers not to fight. These species meet and interact randomly, especially when there is a need to access resources such as food, water, or anything else needed for survival. Having access to these resources improves chances of not only surviving, but also in reproduction. The diagram below shows the average payoff of each species depending on who they get to encounter. A higher number indicates that they can access more resources than the other.

	Hawk	Dove
Hawk	-4, -4	12, 0
Dove	0, 12	6, 6

As you can see, there are three possible interactions for this kind of game—hawk with another hawk, dove with another dove, and hawk with dove. Since hawks are aggressive and are not hesitant to fight for resources, they can end up not only injuring themselves, but also damaging the resource (hence the payoff of -4). On the other hand, the peaceful nature of doves allows them to share their resources (payoff of 6 each). Lastly, a hawk and dove pair results in a mismatch, as the hawk receives everything while the other gets nothing. These interactions also have three possible outcomes—only the hawks will survive, only doves will survive, or both species will survive.

The information above can be considered as signs that things are not good for the doves because of their peaceful nature. However, having an imbalance in nature in that only the hawks will survive is not a good thing. Even if there are periods where the number of doves can go lower because of their inability to access resources, there will also come a time that their population would be able to recover. If you look at it logically, hawks become injured when they fight against each other, and along with that would come the destruction of some resources. If they become unable to gather resources (and also don't share them), there will be an eventual decline in their population.

Going to the side of the doves, since they share resources during situations where they have access to them, they are able to prolong their lives and improve their chances of reproducing and eventually add to the population. Now, since hawks will have less competition, there is more opportunity for them to get more resources because it's more likely that they will encounter doves rather than hawks. This continues until they are able to reduce the number of doves and improve the number of hawks again. Then it cycles back.

This kind of game shows that in order for both species to coexist, one should find the right ratio of hawks and doves. Game theorists have figured out a formula that will allow them to compute this ratio.

These are the steps for this computation:

1. First, one needs to assume that half of the population are doves and the other half are hawks.

2. Determine the average payoff of each species. Since there is a ½ chance that they will encounter an animal of the same species, this number is considered as the constant. Next, multiply ½ with the corresponding payoff of each animal if they encounter another. For example, hawks should get ½ multiplied by -4 (if they encounter another hawk). Do this again for their encounter with the doves, and do both processes for each animal.

3. Once the factors are multiplied, add those numbers to get their average payoff. Given the numbers above, it should look like this:

Hawks: (½ x -4) + (½ x 12) = 4

Doves: (½ x 0) + (½ x 6) = 3

The higher number in this computation implies that hawks will have a higher chance of survival and reproducing, since their average payoff is higher that of the doves. This also means that they should have a higher population as compared to doves.

Although this computation appears to be more of a use to biologists who subscribe to the evolutionary standpoint, it also has some use in psychology. In general, this can highlight why it is not advisable for one to just have a single extreme characteristic. If one wants to continue their survival and even thrive, one should have the perfect balance of traits that may be needed in a situation. One can become hawkish in many instances, but one can also be gentle-natured like the dove in some situations.

Game theory in political science

The field of politics can also be explained or influenced using game theory. Here are some topics in political science that are heavily influenced by game theory.

- **The stability of any political government** - it is believed that game theory can explain how it is possible for the government to remain stable even if the people being ruled upon have higher numbers. Let's apply this to a monarchical government. Although it is the king who is in control of their place, it is not explained as such. Rather, it is explained in a way that their place has a government and its powers are being carried out by the king. This institution also requires subjects or citizens to follow what the government wants and bars any discussion related to replacing people in the institution. The king has given himself a face that may be too large for people to take down (government), so that everyone will remain obedient.

- **Guide the actions of politicians so that they can influence election results** - jumping into the modern times, most countries rely on democracy to choose who can exercise the powers that the government has. Those who want to get hold of that power will have to win the election first, though. With the help of game theory concepts such as the prisoner's dilemma, one is able to strategize when it comes to areas where they should campaign and have a better chance of swinging the results of the election in their favor. This allows them to get the best results with minimal effort, or still get a good payoff regardless of what their competitors decide to do.

- **Decision to go to war** - even in the modern times, it is still possible that countries can wage war against one another even if both parties are aware of all the adverse effects that it brings. War can be caused by certain things related to game theory. For example, if countries or groups (in case of internal issues) cannot fully commit to all the terms in the agreement that both parties are drafting, it can influence them to take arms instead, even if an initial effort to resolve those differences has been made.

- **Territory claim issues** - countries fighting over territories or exclusive economic zones are also influenced by game theory. This kind of issue is dynamic in nature, as the action of one country pertaining to a claim will greatly influence what the other party should do to get the greatest payoff. For example, how one country deals with intruders will surely influence how the intruder reacts and what they will do next.

- **Predicts responses when a new rule or law will be implemented** - changes often shake people, regardless of them being small or big. With political science though, the changes are usually done on a much wider scale and can affect millions of people. Game theory can analyze how people will react to a proposed change so that those in authority can better prepare for possible questions, and patch up holes before the new law is implemented.

The game of trespass

A good game that is somewhat similar to territorial issues is trespass.

It is a sequential game between a landowner and a hiker. Let's just name them Charles and Megan respectively. Charles owns some land close to the river that is part of the countryside. Megan, on the other hand, usually hikes on the countryside and would prefer to pass through the property of Charles, rather than walk around and not trespass on his property. In order to deter Megan from going through his property, he made the first move by putting up a sign to warn anyone who is planning to trespass on his property that they will be prosecuted. Megan eventually sees this and should decide what her action will be.

For this game, Megan can choose between two choices:

- Megan ignores the warning and goes through the property anyway.

- Megan chooses to not go through the property and avoid any chance of getting prosecuted.

To analyze this situation, one should look at the benefits and drawbacks that each party will get, which are as follows:

- If Megan decides to go through even if there is a warning sign, she benefits from by getting to her destination at a much faster rate. Charles, on the other hand, will be forced to make another move and decide if he should follow through on his threat or not.

- If Megan decides to heed the warning, she'll be inconvenienced with the detour. With this action, the game ends, since Charles will have nothing to do.

The length of this game depends on whether one complies with the previous move or not. This is different from other dynamic games wherein participants only get to move once and get their payoffs after each of those moves.

If Charles decides to prosecute, the game can drag on and even result in the involvement of another party to settle the issue.

This somehow emulates territorial disputes. If the request to get off the territory is disregarded by the other country, the game drags on and can even come to a point wherein the intervention of a bigger organization that has the ability to make one of the parties comply will be needed.

Uses in technology

The advancement in technology made it possible for people to make machines capable of providing a reaction based on their user's action. Think of modern computers as being taught with complex decision trees, or have multiple payoff matrices programmed in them so that they can act accordingly based on the user's input. It's even possible to converse with AI and get a response that looks like it came from an actual person.

It also has allowed us to automate certain tasks that may be too complex if done manually. For example, it's now possible to verify if the format of an email address is valid in a matter of seconds; because programmers are able to "teach" computers how to do the job and execute another command if one fails. One can input thousands of email addresses and get results immediately—something which would be time-consuming if done manually.

Chapter 4 – The Prisoner's Dilemma

One popular concept in the study of game theory is the prisoner's dilemma. This chapter will provide an in-depth explanation of this concept so that you can further understand game theory.

What is the prisoner's dilemma?
This refers to the standard situation commonly analyzed in game theory and decisional analysis. This paradox shows that two individuals usually act for their own self-interests, resulting in an outcome that deviates from what is considered to be ideal.

This concept is set up in a way that if one chooses to protect themselves, they do so at the cost of them either being harmed, or causing a detrimental result to the other participant.

Simply put, this concept shows that when given a choice, self-interest is usually more desirable. However, choosing self-interest could result in a much worse outcome, especially if both parties chose to take an action based on their own self-interest.

Applying the prisoner's dilemma
To further understand the concept, it's necessary to study a scenario on how it is applied.

Suppose that there are two people. Let's just call them James and Andy. They were arrested because of a crime that they allegedly committed, and are now brought to two separate rooms for interrogation.

Now, both of them will surely want to minimize their jail time. But in order for this to happen, they must make their individual decisions based on the consequences that are provided to them. Both of them are informed that they will get one of the following consequences depending on what their partner has decided to do:

1. If one of them has decided to confess to the crime and the other remained silent, one will be given a three-year sentence while the other is set free.

2. If both of them chose not to confess, they will each be imprisoned for a year.

3. If both of them confess or show evidence about the crime and each other, they will each spend two years in prison.

Given these consequences, it's easy to see that the ideal situation so that they only get the minimum sentence is for both of them to remain silent about the crime.

However, the dilemma here lies in the "possibility" of the other participant to cooperate and think of not confessing as well; after all, it's not possible for them to think about the course of action that should be taken. There is also the possibility that one will not be convicted of the crime, allowing them to go free in exchange of the other staying in prison for three years. In situations like this, self-interest strongly motivates one to choose the option that will allow them to go free even at the expense of the other participant being sent to prison.

Upon analyzing the outcomes of each action, both parties will surely confess things about their partner, even if it is obvious that they'll get the minimum sentence by not saying anything. This is because regardless of the action of the other party, they will surely get a better (or at least a fair) result. It's better to stay in prison for a little bit longer with their partner, rather than see their partner free while they are imprisoned. Ultimately, it can be said that the dominant strategy for this game is to confess or betray the other, as taking another action will give them a disadvantage.

How is it applied in other fields?

Even if this concept makes use of potential prisoners, it can also be applied in real-life situations. Here are some areas where this concept can be seen:

Psychology

People who want to get rid of their addiction can also experience the prisoner's dilemma. In this context though, confessing is the same as relapsing (going back to one's old behavior).

However, relapsing can serve one's self-interest because they are able to consume their object of addiction, which has a strong influence on their behavior.

If one is to follow the logic of the prisoner's dilemma, they have to face four different outcomes:

1. Not relapsing both today and in the future is the ideal situation, as this is a sign that the person is able to break free from their addiction even without close monitoring.

2. Not relapsing today but doing so in the future is considered the worst outcome. Doing this only means that the person will have to start from scratch so that they can recover from their addiction.

3. Relapsing today and in the future is considered to be slightly better. Since the person has not started any form of intervention yet, no effort has been wasted.

4. Relapsing today but not in the future is similar to getting one more day or one last chance to be in contact with their object of addiction before they start making an effort to get rid of it.

If one is to analyze those options, the last option is usually what most addicts experience. They want to change their behavior, but reason that they can't instantly change. This influences them to put off their attempts in changing their behavior for another day. However, when tomorrow comes, they have to face that same dilemma. If they are unable to break this pattern and repeatedly choose that same option, they are surely trapped in their addiction. It's as if it is better for them to continue with their addiction than to see their efforts fail, and go back to square one if they become unsuccessful in resisting temptation.

This same situation is also true if we are to go back to the prisoner's dilemma. Even if a person knows that the best result will be experienced by choosing not to confess, they will always choose to confess so that they will not get the short end of the stick in case the other participant chooses to betray them. In this context, they are stuck with choosing the same decision.

Economics

Two situations in economics where the idea of prisoner's dilemma can be seen are advertising, and determining the price of goods.

Suppose that there are two or more makers of a certain product and they would want to improve their sales. To make this possible, they would have to spend money for advertising. However, it's highly possible that the advertisement will be seen by their competitors. Since advertising will give the benefit of getting recognized and the potential of higher earnings, if one company starts doing this, others will surely follow. This is because not following on the trend will only place them in a disadvantageous position. Even if their advertisement will just cancel out the effect of other company's advertising, they'll still do it just so their rivals will not get ahead and earn more.

The same concept is applied when deciding the price of goods. Companies can choose to cooperate and follow the suggested retail price of goods so that everyone is on the same level. However, they can also betray their competitors and offer their goods at a much lower price, especially if they are willing to slash off their potential profit in the hopes of enticing customers to buy from them.

If their competitors see what they are doing, those companies will surely lower their prices so that they can also have a share on the customers who are availing similar goods for that price. In a way, it forces companies to act in a certain way just so their competitors will not take away the customers that they previously acquired. This trend of adjusting the prices can continue until such time that no one is willing to adjust it anymore.

The prisoner's dilemma is observed in fields other than what is discussed above. While it is easy to see what the ideal solution is, it makes use of the influence of self-interest in the decision of the participants. All of these decisions, although not ideal, are rational.

Chapter 5 – Cooperative Games and the Shapley Value

There are situations that require people or entities to work together in order to experience some form of benefit. This is what cooperative games are. However, when it's time to get those benefits, how can one determine how much a group should get?

Arguably, this is the most applicable and relatable among the things we have discussed so far. After all, if you're currently attending school or if your job involves collaboration (which it probably does), you will realize ways of making work much more efficient and rewarding.

This chapter will provide additional information about cooperative games and how you can determine fair distribution using the Shapley value.

The presence of social loafing

Although it is obvious that cooperation brings benefits to its members, one primary problem observed in any group setting is social loafing.

Social loafing is a concept in social psychology which describes how there is a tendency for a person or entity to exert less effort if they are working in a group rather than if they work alone.

If this situation does arise in many groups, and members are less interested in exerting their best effort, then there is a possibility that groups can be less productive compared to doing work individually.

Being part of the group also produces a single product. With this kind of setup, it's highly possible that not everyone will get credited for the whole product. This can affect one's motivation to act for the purpose of helping the group, as recognition is not always given.

Why cooperate then?

If social loafing is indeed observed in some groups, why do some participants still engage in cooperation?

The first reason for doing this is because of the advantages in streamlining a process. If there's a need to get results faster, it's easier for a group to create and implement a process that will make it possible. For example, if someone wants to establish a business of baking cookies with his/her friends, it would be best to assign one participant to kneading the dough, another to forming the cookies in shape, and other members for the remaining processes. Since each participant is focused on doing one specific task, it's easier to pick up the pace and get better at it. Eventually, it will result in faster production, as members are not required to fulfil one task after another just to create the same product.

Another reason is to achieve a goal that may be unreachable for a single participant. Using the same example as above, if the participants are given a small amount of time to achieve a huge number of cookies, it would be in their best interest to help one another so that things are done faster. It's just like hiring other people to do your work so that you can work on other, equally important tasks.

What is the Shapley value?

Once the coalition is able to achieve the goal that they have in mind, it's time to get their payoff. This is where the Shapley value comes in, as it aims to provide a model that can be used to compute how the total payoff should be divided among the participants.

The Shapley value follows these axioms:

1. Determines a player's marginal contribution - a player's marginal contribution to the group is determined by knowing how much the group is gaining or losing if that player is removed from a game. For example, let's say you have a business that can produce 100 pieces of clothes in a day. If one of your employees needed to be absent and the production is suddenly reduced to 80 for that day, their marginal contribution to the group is 20.

2. If there are two or more players that bring the same kind of contribution to the group, their targets as well as their payoffs should be the same - there are instances where two or more people are needed to do the same task. Since the requirements for the job is the same, the outputs expected from them should also be the same.

3. A participant that contributes nothing to the group, then they should receive nothing - even if a player is part of a coalition, not doing anything for it also means that the group should not give them anything. In game theory, these people are referred to as dummy players.

4. If the game has more than one part, then payment or cost is determined by their contributions for each part. For example, if a player has contributed a lot on one day and not so much on the next, the payment for each of those days should be different.

Now, let's say you and your friend wants to start a business of baking and selling cookies. Since you'll be teaming up with another person, it's highly possible that you'll be able to work faster because the process is now more efficient (one mixes all the ingredients while the other is tasked to monitor the baking). Your collective effort can produce 50 cookies in an hour. The difference is significant compared to what each of you can do individually (say your friend can only make 15 cookies while you can make 25, giving the combined total of 40 in an hour). Eventually, all cookies are sold for a dollar each, giving the coalition 50 dollars.

The question now is: *how will the income be split?*

This is where determining the marginal computation of each person is important.

First, we take a look at you and your friend's marginal contribution based on your possible individual efforts. To do this, subtract individual contributions from the total combined number of products. Given the figures above, if 50 is the total number of products, 25 and 15 should be subtracted. This will give you the numbers 25 and 35, respectively. The first number indicates your friend's marginal contribution to the group, while the other one is yours.

Since the marginal contribution of each party is different, those values should be added and averaged. With the average of 30, it is then decided that the person who has a higher marginal contribution will get that amount while the remaining 20 dollars will go to the other. The average is known as the Shapley value, as it determines how the payoff should be split. You can make use of this model no matter how many participants, or regardless of the amount of money earned.

If the same process is done on the next day, another computation should be made, as there is a possibility that the number of products made and the profit generated might be different.

The application of other axioms

In the example above, the axiom of the Shapley value that was mostly used is the marginal contribution. Looking at the situation though, we can see that it also applied the axiom wherein the production every single day should always be taken note of, so that the pay of each member for each day depends on the effort that they have exerted. If you slacked off on the next day and produced less than yesterday, it's only fair that you will receive a much lower payment as well. Of course, if someone can't work for that day, that person will get nothing because their marginal contribution for that day is zero (following axiom 3).

Axiom 2 is also applied in that situation. Since each does the same work (that is, under the production of cookies), the computation of their payoff should be computed just the same.

Other points of discussion

It's easy to see that the Shapley value is directly used in the field of business, especially when it comes to determining how much an employee should be paid based on the number of products that they can produce. However, there are other points about the axioms that needs further discussion.

1. **Salary differences between entry-level and experienced employees** - taking a look at axiom 2, it mentioned that one should give the same target and payoff to people who provide the same contribution to the group. While the target or quota part is almost always met, there are times when there is a difference between the payoffs provided to different people even if they hold the same position. They may provide the same contribution to the company, but the salaries consider the player's work experience as well.

2. **Payoff can also be given to dummy** players - it's ideal that those who did nothing for the coalition should not receive anything. However, this is not always the case for modern companies. There may be reasons as to why a member cannot contribute to the coalition, such as if they get into a work-related accident, or the participant might

need to go on maternity leave. While these kinds of participants are unable to contribute, the group can still provide payment to them.

3. **Some works cannot be quantified** - looking at the modern field of work, there are jobs that may be difficult to quantify. For example, human resource employees may be asked to make policies that will be beneficial, but they cannot be paid based on the number of policies that they can make per day. Consider what the job entails, as it is not always possible to put a number on what the player has contributed to the group.

Chapter 6 – Game Theory in Oligopoly

It was mentioned in the previous chapters that game theory is heavily applied in the field of economics. One such area that fully uses game theory is oligopoly.

This chapter will discuss what this market situation is, and how it uses game theory.

What is oligopoly?

This term refers to a market structure wherein the number of firms who are part of it are quite small, but not too small that they are considered as monopolies. While there is no upper limit as to the number of companies in that industry, the number should be small enough that an action of one firm can significantly influence others. Think of oligopolies as a game where you need to compete with a few people. Since the number of participants isn't that huge, any action that one takes is easily noticed by others.

An industry where entry isn't that easy can be part of an oligopoly. For example, wireless carrier companies are part of an oligopoly since there are a limited number of service providers. At the same time, it's also not easy to penetrate this industry because of the cost that it entails to start and operate this kind of business. The same can be said for companies in the fields of cable television, mass media, and vehicle manufacturing.

Here's a more specific example—operating systems. If you have been using a smartphone these past few years, you probably know that there are two main options when it comes to mobile operating systems, Google's Android and Apple's iOS.

Important points in oligopoly

All companies who are part of this market structure are mindful of these important points:

Set prices

Since oligopoly firms have relatively few competitors, they have some control over the price of what they offer; after all, potential customers don't have too many choices. Think of this as a scaled-up application of the law on demand and supply.

But even if companies in an oligopoly market have the liberty to do this, they don't change the price of their goods at will just so they can earn a lot and maximize their profits. In fact, it's less likely for them to support or implement a price hike, and they are more willing to implement a price change if it is a reduction. Once they have decided on a certain price, the product or service usually becomes fixed to that price, and it remains as such even if costs change for the company or the products.

Non-price competition is the name of the game

Since changing the prices doesn't always benefit firms who are part of an oligopoly, how can they entice customers to choose their products then? The answer would be to use non-price tactics.

Certain actions can influence customers to prefer one brand over the other, even if the price of what is being offered is the same for companies.

Here are some of the techniques that firms in an oligopoly market use:

- **Innovating their product** - products can be improved so that customers are more likely to buy them. The evolution of cellular phones to become what is now known as smartphones is a great example of product innovation.

- **Loyalty schemes** - some customers consider what they can get from buying a certain brand. One benefit of doing this is to introduce a loyalty program wherein customers get extras if they fit the company's criteria of who can be considered as a loyal customer.

- **Quality of service, including aftersales** - there are customers who want to feel that the company isn't only after their money, and that the good treatment should not end when the customer steps out the store. This explains why they have a service center or office where customers can head to in case there are concerns after the purchase. For example, those who sell smartphones or computers have technicians that can take a look at their customer's device in case something is wrong. It may not be free, but it's something that buyers strongly consider before buying a product.

- **Branding** - the products sold by firms in an oligopoly market for a specific industry are somewhat similar. For example, Dell and HP both sell computers. Since this is the case, each firm must show something unique in their product that their customers care about. Doing this makes the brand known for that particular feature or edge, making it easier for them to entice customers who are looking for those features.

- **Freebies** - customers are more enticed to buy products from a certain brand if they can get free items or additional benefits without having to pay extra.

There is possibility for collusion
Collusion happens when different companies decide to work together so that the benefits that they are getting from their current situation are improved.

Since there are few companies in an oligopoly, it's possible for them to work together even if they still exist as different entities.

Firms in an oligopoly market that have decided to collude are called a cartel. They recognize that their companies are mutually interdependent, and they will maximize their profit if they work together.

Collusion might seem fishy because it looks like companies work together so that they can get the better of their consumers. However, there are two advantages that can be experienced in colluding:

- **Reduces the cost of competition** - it was mentioned in a previous chapter that companies can choose to advertise their products or services so that they can entice more customers. While this can also be done by firms in an oligopoly market, they allocate less money to it. Since they have one or a few allies in the same industry that help them get what they want, there's no need for them to spend money just to highlight why their product is better than those of others.

- **Can reduce uncertainty in the market** - companies who work together can earn maximum profit and this can often result in a surplus. For the buyer, this only means that there is less likelihood of a product shortage. Given this case, prices will likely remain the same or can even go lower.

Applying game theory

Game theory is applicable because oligopoly markets have a relatively low number of competitors, making it a perfect information game.

The situation is also similar to a sequential game. In many oligopolies, there are companies which are considered to be the price leader. If they have decided to set the price of a product as X dollars, other companies will surely react accordingly. Although other companies rarely deviate from that price, this situation looks like they are waiting for Player 1 to move before they think and act based on the action that was taken.

Collusion is also similar to a cooperative game, since two or more companies may work together so that they can maximize their gains. However, collusion can also breed competition that is similar to a prisoner's dilemma.

Collusions can still be fragile, and there is no guarantee that cooperation can be expected. A company that is part of a cartel can offer their products at a lower price (similar to confessing or betraying the other), even if a previous discussion enforces others to offer at a high price. If they do so, customers will surely notice the difference and take advantage, and the company can get most of their competitor's customers to buy from them.

Why place special emphasis on oligopoly?
While there are other market structures and all of those can also make use of concepts in game theory, oligopoly is special because of some important characteristics.

Firms in other market structures can also have price leaders that dictate how their goods or services should be priced. But even if this is the case, they still have variable prices for the same products. Oligopolies also make use of price leadership; however, they don't change prices at will, even if they can do so. Based on this analysis, it's safe to say that prices in an oligopoly market are more stable compared to firms that belong to other market structures.

Another point in oligopoly is the common trend in using non-price tactics. Not all market structures are able to employ those tactics. For example, those who sell identical items such as fruits or veggies never use advertisements because it will simply drive costs up and affect their price. Since it's easy for customers to simply go to other vendors who sell those goods for a lower price, it would be best for vendors of those kinds of goods to not spend too much on advertising. Companies in an oligopoly though, have the freedom to do this. Even if these firms spend on advertising or use other non-price tactics, customers will not see any change in the price of the items.

The presence of collusion is also different for oligopolies. Firms outside the oligopoly do not often cooperate with one another; usually, companies will only work in unison with others if one company is merged into another. In oligopolies, there are incentives if they work together. This is not the case for firms outside this market structure, as they are mostly competing with others.

Entry deterrence game

Since there is no upper limit to the number of companies in an oligopoly market structure, there is a possibility that a firm will enter a particular oligopoly industry and become one of its players. For example, Company X can enter the mobile phone industry and offer new models for consumers, or Company Z can expand and venture as a new player in the computer industry. However, the entry of another player in the industry can affect other firms who are already part of it. Think of it as introducing another fish in a pond that may already be too small for those in it.

For firms of that small "pond", it would be best if they can deter competition from entering or drive them away once they are in the scene. Some of the tactics that they can use to make this happen are as follows:

- **Threats of an expensive advertising war** - companies in this market structure have built their resources while being in the business for years. Since the entrant is still new, it's highly possible that they do not have enough money to spend on advertising. If the older firms decided to advertise, it's likely that consumers will overlook the new player, preventing them from getting a share of the market.

- **Maintain output** - one possible explanation as to why some firms decide to enter the oligopoly market is that there may be a gap in the current supply or quality of the product sold by the current players. This can be the opportunity that they are waiting for to penetrate the market and share it with the big players. However, those who are part of that industry have the means to improve or maintain their current output so that consumers will choose to stay with their products. If successful, the new player is again left with little to no potential customers.

- **Predatory pricing** - members of the oligopoly can also act by lowering their prices once the new player enters the scene. This will force the entrant to also adjust their prices to match those of others, and entice consumers to try out their brand. The question though is how low the price will be adjusted. If the price is too low, the new player might not make a profit.

Entry deterrence is the closest possible situation in game theory that resembles the scenario of having another player in their industry. In this game, the participants would be the members of the oligopoly, plus the potential new player. This is a sequential game in which the new player gets to act first.

To easily understand this situation, it would be best to assign numerical values for each outcome. These are as follows:

- Assume that the value of the market is worth 10 for the members of the oligopoly. If the new player is allowed to enter the market and get a share of the customers, the value of the market for the players will become 5.

- If the new player decides to enter the market but the other firms will employ the tactics mentioned above so that the new player will not last in the industry for long, both parties get a -1 value, signifying that it will result in a loss of profit.

- If the new player decides not to enter the scene, the members of the oligopoly will do nothing. The initial value of 10 will be retained to them, while the potential new player will neither gain nor lose anything.

- In this situation, it can be seen that the members of the oligopoly have two possible strategies to use; either let the new player enter the market or fight it.

- Based on the numerical values, it can be seen that the best course of action to take for both parties would be for the new player to enter the scene while the existing players share the market with them.

What about the tactics then?

Given the values above, it is shown that fighting the entry of the new player and using predatory pricing or the other tactics can cause a loss to existing players. This raises the question of when should these tactics be used?

Although the tactics that existing oligopoly players can use to deter new entrants or drive them out of the market can be effective, they should only choose to fight if these conditions are met:

- If the result of the effort that they can get from fighting will exceed their potential gain from allowing the new player to enter.

- If the value of their effort subtracted from their potential gains if the new player is not allowed to enter the market is greater than what they can get if they chose not to fight.

To represent the first condition, first look at the values provided above. Doing that gives this formula:

$$\text{If } -1 + d > 5 - c, \text{ fight new player's entry}$$

The value of -1+d refers to the payoff that the existing players will get, from deterring the new player, while the value of 5-c is the payoff that they can get from conceding. Values should be added or subtracted if existing players choose to make an effort against the new player. While they are represented as different letters, the values of c and d are the same (since these refer to the numerical value assigned to the effort that they made).

Let's say that the existing players chose to participate in doing social works to improve their image to the public, and possibly get customer loyalty even before the new player has started operating. They have assigned this effort with a value of 5. Apply this value with the formula above and it will result in the values of 4 and 0 respectively. Since this course of action will give them a payoff of 4 without an additional competitor, there is a benefit to fight the entry.

The purpose of the second condition is to know if it's worth paying the cost of deterring the new player's entry. If one is to use a formula for this, it will be this:

10-c > 5

10 is the value without a new player and 5 is its counterpart. Again, we need to assign a numerical value to c, which is the cost or effort made by the existing player.

Using the same value above, subtracting 5 from 10 is 5. Since this is not greater than the payoff if the new player is allowed to enter, the effort to be exerted is not enough to cover the cost that they will have to pay for the deterrence.

In this situation, since one of the conditions are not met, it's not advisable for the existing players to deter the new one. Again, the former should only take this action if they can make an effort that will bring them a positive payoff for both conditions.

Conclusion

Thanks again for choosing this book!

I hope you enjoyed learning about game theory! As you can see, game theory can be applied to many different areas of life, and business. Use this to your advantage, and implement game theory in your daily life to make better decisions!

If you enjoyed this book, please take the time to leave me a review on Amazon. I appreciate your honest feedback, and it really helps me to continue producing high quality books.